D1384535

# SEAWATER
## A Delicate Balance

## A. Lee Meyerson

— an Earth Processes book —

ENSLOW PUBLISHERS, INC.

Bloy St. & Ramsey Ave.    P.O. Box 38
Box 777    Aldershot
Hillside, N.J. 07205    Hants GU12 6BP
U.S.A.    U.K.

**Library of Congress Cataloging-in-Publication Data**

Meyerson, A. Lee (Arthur Lee), 1938-
    Seawater: a delicate balance / by A. Lee Meyerson.
        p. cm. -- (An Earth processes book)
    Summary: Describes the chemical composition, or saltiness, of the
ocean and discusses the effect of man on the oceans as well as the
processes that formed the oceans and now maintain their balance.
    Includes index.
    ISBN 0-89490-157-5
    1. Sea-water--Composition--Juvenile literature.
[1. Oceanography.] I. Title. II. Series.
GC101.2.M48 1988
551.46--dc19

                                                            88-10961
                                                                 CIP
                                                                  AC


Printed in the United States of America

10 9 8 7 6 5 4 3 2

**ILLUSTRATION CREDITS:**
*Alvin* external camera, Woods Hole Oceanographic Institution, p. 34;
T. Cosmas, pp. 18, 19, 36, 38; S.C. Delaney/Environmental Protection
Agency, p. 8; Dudley Foster, Woods Hole Oceanographic Institute, p. 45;
Kitt Peak National Observatory 4-meter (157-inch) Mayall Telescope
Photograph, National Optical Astronomy Observatories, p. 22; Lamont-
Doherty Geological Observatory of Columbia University, pp. 27, 28, 29,
32, 42, 51; (MSH-Hoblitt, R.P. 13), U.S. Geological Survey, p. 50; John
Porteous, Woods Hole Oceanographic Institution, p. 49; U.S. Geological
Survey, pp. 25, 41; Woods Hole Oceanographic Institution, p. 44.

# Contents

# 1

# Mankind and the Sea

*March, 2086.* The beach is crowded with college students taking a break from their studies during the spring vacation. The waves are gently crashing on the shore beneath the glare of the hot sun. It is difficult to walk along the beach without stepping on a sunbather. As the lifeguard gazes out over the water, he sees people everywhere. It is going to be a long day. It will be hard enough to *find* people who are in trouble in the water, let alone to *save* them. If all these college students wouldn't come to the New Jersey shore for their spring break, life would be a lot easier.

Spring break in New Jersey, not in Florida? No, that is not a mistake. It is possible. Without intending to, mankind is now conducting the largest experiment in the history of the earth. By producing huge amounts of energy, we are adding gases and smoke particles to the atmosphere, and we are not sure how they will affect the planet.

The gases that are being added to the atmosphere can prevent heat from the earth from being lost to outer space. One of the gases added to the atmosphere is carbon dioxide. This gas and others are producing an effect that is much like that of a greenhouse. The heat from the sun can pass through the atmosphere and warm the earth, but not all of that heat can be returned to space. The result is that the average temperature of the atmosphere will rise. Most scientists are sure the temperature of the atmosphere will go up, but they don't know by how much.

The amount of warming may depend upon the way the ocean works. The major gases that are causing the problem tend to dissolve in the ocean. At present we do not know how much of the added gases will be absorbed by the ocean and how much will remain in the atmosphere. When that question is answered, we will have a better idea of exactly how warm the atmosphere might become.

As the atmospheric temperature rises, the climatic zones will shift to the north in the northern hemisphere and south in

A beach in New Jersey. Perhaps mankind's great experiment with the atmosphere will cause this beach to become more popular in early spring.

the southern hemisphere. Perhaps the shift northward will be enough to allow swimming in New Jersey in March. If the atmosphere is warmer, more storms may occur and they may be more intense than they are now. If storm patterns change and climates shift, the entire pattern of agriculture may change. Areas that are good for farming today may not be cropland due to the temperature increases and changes in rainfall. With sufficient atmospheric warming, the great agricultural area of the midwestern United States may no longer be able to feed the country. That excellent farming climate may shift northward to Canada.

This great experiment is underway, and we may see the results by the year 2000 or shortly afterwards. We can only guess what they will be from what we already know about the workings of the system we call earth. We can delay or prevent atmospheric warming by decreasing or stopping the release of gases into the atmosphere. However, the gases are released by energy-producing activities. When we heat our houses, drive our cars, and use our refrigerators, we are adding heat-retaining gases to the atmosphere. We can delay or prevent this process but are we willing to give up these conveniences? Probably not. So the experiment will continue unless we can find alternate ways to produce energy, ways that do not release heat-absorbing gases.

The increase in the temperature of the atmosphere will not be the only effect of this great experiment. The ocean will also react to atmospheric warming. As the temperature of the atmosphere rises, water now locked up as ice in glaciers, on Greenland and Antarctica, for instance, will be returned to the sea. In addition, the water of the ocean will expand with rising temperatures. The result will be an increase in the volume of water in the ocean, and consequently a landward movement of the shoreline as the water level rises.

As the sea level inches upward, coastal areas will come under increased attack. Beach erosion will proceed at a faster rate. The more intense storms will cause even greater erosion and coastal flooding. All coastal towns are now faced with a decision that will determine their future. How must they react to this predicted change in the shoreline? No matter what they do, the changes that will occur may be significant, either due to a rise in sea level or due to structures built by the towns.

There may be many other changes in the ocean that are not as evident as the rise in water level and erosion of the coast. If some of the added atmospheric gases go into the ocean, the chemistry of the ocean must be affected. This is a possibility we must consider. Will the addition of one substance to the ocean affect others? The ocean appears to be a complicated system in which one change brings about others.

Mankind is adding enough gases to the atmosphere to affect the climates of the world.

If its chemistry is affected, will that affect mankind? These questions can be answered only if we truly understand how the ocean works.

Another experiment that is going on has immediate effects on the well-being of the ocean, its inhabitants, and mankind itself. Historically, we have used the ocean as a dumping ground. Whenever they needed to dispose of waste products, those who had convenient access to the ocean merely flushed them out to sea. The ocean was thought to be so vast that it could absorb anything that was dropped into it.

In the United States alone, approximately 50 million tons of waste materials are dumped into the ocean each year. If that were not bad enough, those who dump do not take their cargo out to the middle of the ocean. The wastes are dropped in the waters of the continental shelf, usually less than 18 miles (30 kilometers) from the shore. The impact of dumping in such relatively shallow waters is far greater than it would be if it occurred 600 miles (1,000 kilometers) out.

The oceanic area that receives more wastes than any other in the United States is the continental shelf off the coast of New Jersey and New York. Environmentalists refer to this location as the Dead Sea. Few organisms live on the seafloor in that vicinity. Dumping takes place close enough to shore to contaminate the bathing beaches. Naturally, organisms that spend any length of time in the site of the dumps will be affected.

Just as the ocean, as a whole, is a complicated system in which everything depends on everything else, so too are the inhabitants of the ocean. Some organisms have the ability to live in polluted waters. However, they tend to concentrate in their tissues certain chemicals that can be toxic when consumed by other organisms. The result for the consuming organisms can be disease and death. Can this affect mankind?

*1953. Minamata Bay, Japan.* A few people began to show up in the hospital emergency room with strange physical and mental problems. Animals were found dying a convulsive death in the streets. The cause was a mystery. Since all the people had the same symptoms, they had to have something in common. Eventually it was found. The people and animals had eaten shellfish from the bay. The shellfish were examined and found to contain high levels of mercury in their tissues. Eventually one hundred people displayed the symptoms; almost half of them died.

The cause of the poisoning was the dumping of mercury-

The sea, a vast and complex water solution that is the last great frontier for exploration of the earth.

containing wastes in the bay where the shellfish lived and fed. Of particular interest and importance, the dumping began in 1938. It took fifteen years for the effects to show up in people. Just because we do not immediately see effects when dumping occurs it does not mean that harm is not being done. The system of the ocean works slowly, but it does work.

The ocean is not merely a vast pool of water filling up a great depression in the earth. It is a dynamic system that we have the ability to change. Should we continue to use the ocean as a great experiment without fully understanding the consequences? Perhaps a better idea may be to attempt to find out how the ocean works. We already know some things about the ocean system, but certainly not everything. The ocean is the last great frontier on earth. We know less about the depths of the sea than we know about the surface of the moon. The time has come to learn about the ocean.

# 2

# Seawater Is Salty

The Los Angeles airport was crowded with people, all excitedly talking. It was almost impossible to hear oneself think. Everyone was about to board a plane—destination Hawaii. With a trip like that about to begin, it was no wonder that everyone was excited. The noise didn't stop, even when the plane was loaded and began to taxi down the runway.

As the plane lifted, it made a turn and headed over the Pacific Ocean. For the next four hours, all people could see out the windows of the plane was water. It takes as long to fly over the Pacific Ocean from Los Angeles to Hawaii as it does to fly across the entire continental United States. That is a large amount of water to cross, and Hawaii is only halfway across the Pacific Ocean!

If you live in the continental United States, it is difficult to imagine how vast the ocean is. However, on an island, such as one in the state of Hawaii, one never forgets how vast the ocean is and how insignificant is his own little piece of land. People in Hawaii can drive to the top of a volcano, over 10,000 feet (3,000 meters) high, and see that they are surrounded by water. In every direction the land quickly ends. Beyond lies the vastness of the ocean.

A view of the ocean brings to mind many different things. Each person who sees it paints a different picture in his/her mind. For some it is beauty, for others it is sport, for others a barrier to be overcome. It is even more than that to a chemical oceanographer, a person who studies the composition of the ocean. The chemical oceanographer sees not only the water that makes up most of the ocean, but also what is dissolved in it. After all, we know the ocean is salty, but what does that mean, and how did it get that way? These are only two of the questions that need to be answered about the ocean.

Although we usually think of common table salt (sodium chloride, NaCl) when we mention salt, the saltiness of the ocean includes much more. Water has the ability to dissolve many things, a process we can easily see by putting sugar in a glass of water. Everything that is dissolved in the ocean helps to make it salty. The elements in common table salt, sodium (Na) and chlorine (Cl), are the most abundant elements dissolved in the ocean, but all naturally occurring elements that can be dissolved in water are present in the ocean, although most of them are present in only extremely small amounts.

To get an idea how salty the ocean is, we can take a gallon (3.78 liters) container and fill it with water from a faucet. If we add 4.79 ounces (132.39 grams) of salt, it will contain the same proportion of salt as does seawater. That is not very much. If we think of seawater in terms of its mass, 3.5 percent of its mass is composed of substances that are dissolved in the water. The remaining 96.5 percent is water.

The total of all the substances dissolved in a sample of seawater is called salinity. So we can say that the salinity of a typical sample of seawater is 3.5 percent. However, if we think about it, 3.5 percent is a very small amount, especially when it is divided among a large number of elements. We would be talking about most elements being present in quantities of less than 1 percent. To make it easier to deal with these low concentrations, oceanographers change the units slightly. Percent is really the number of parts present in 100 parts, or parts per hundred. If we increase the total number of parts, we can deal with larger numbers. Instead of thinking in terms of 100 parts, let us think in terms of 1,000 parts. In this case the salinity would be 35 parts in 1,000 parts. (All we have to do is multiply percent by 10.) We would then say that the

### SALINITY OF SEAWATER

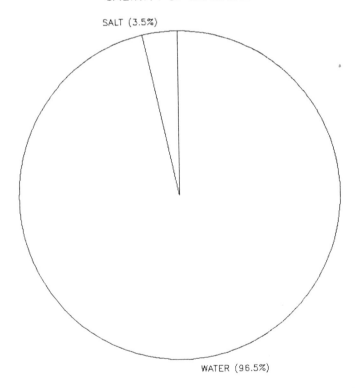

In average seawater, 3.5 percent of the mass consists of dissolved salts.

salinity is 35 parts per thousand, which is written as 35 ‰. The symbol, ‰, comes from the percent symbol, %. Since we multiply percent by 10 to get parts per thousand, we merely add an extra zero to the percent symbol to get the proper symbol for parts per thousand.

The fact that at least 83 elements are dissolved in seawater seems to make its composition very complicated. Actually, it is not as complex as you might think. Only six substances make up 99.28 percent of everything dissolved in seawater. These substances, the major components of seawater, and their concentrations in parts per thousand are as follows:

| Substance | Parts per Thousand (‰) |
|-----------|------------------------|
| Chlorine (Cl) | 19.35 |
| Sodium (Na) | 10.76 |
| Sulfate ($SO_4$) | 2.71 |
| Magnesium (Mg) | 1.29 |
| Calcium (Ca) | 0.41 |
| Potassium (K) | 0.40 |
| | |
| Total | 34.92 |

Notice that out of 35 ‰, only 0.08 ‰ is left for everything other than the six major components. Most elements that are present in seawater exist in very small concentrations.

Now that we know what salinity is and which substances make up the major components of salinity, we should know how to measure salinity. This is difficult to determine directly. We could take some seawater, evaporate it, and measure what is left. This, however, is hard to do accurately, since some of the dissolved materials are gaseous and can be lost during evaporation. Fortunately, there is a solution to this problem.

In 1845, Johann Forchhammer, a professor of geology in Copenhagen, Denmark, decided to learn as much as possible about the composition of seawater. He wanted to collect and analyze samples from all over the world, which was a difficult task, considering how slowly the sailing ships of that time moved. A single voyage could take many years and visit only a small portion of the world's oceans. Therefore, rather than collect water samples himself, Forchhammer asked sailors to bring him water samples from wherever they went on their cruises. In that way, he could collect a great number of samples from a great number of different locations.

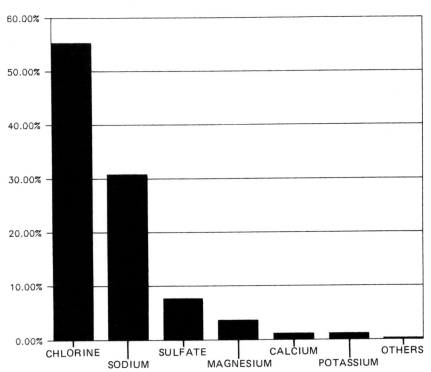

MAJOR COMPONENTS OF SEA SALT

A total of only six components make up over 99 percent of the dissolved salts in average seawater.

For twenty years Forchhammer collected and analyzed seawater samples brought to him by sailors, a true example of scientific dedication. When he published the results of all that work, in 1865, he demonstrated that although the total salinity of seawater differs from one place to another, the ratio of the major components to one another is always the same. For example, no matter what the salinity of the water in the open ocean is, if we divide the concentration of sodium by the concentration of chlorine we will always get 0.55. If we divide the concentration of magnesium by the concentration of chlorine we will always get 0.067. Even today, with all the advanced instruments that enable us to determine the composition of seawater to a greater degree of accuracy, we find that Forchhammer was correct.

Based on Forchhammer's discovery, Martin Knudsen, a Danish oceanographer, devised a method for determining salinity merely by measuring the chlorine concentration, or chlorinity. Knudsen reasoned that if he could find the concentration of one major component, Forchhammer's ratios would allow him to calculate all the others. Chlorinity is easy to determine accurately and, therefore, was the best major component to use for this indirect method of determining salinity. Knudsen then devised a simple relationship that allowed him to calculate salinity from the chlorinity measurement. His measurements were so accurate that a standard was developed to which all measurements could be compared. Called standard seawater, it is now sold by several oceanographic laboratories for use all over the world. It consists of ocean water that has been analyzed for chlorinity to the nearest ten-thousandth of a part per thousand, and it can be used to ensure that all instruments that measure salinity will be uniformly accurate.

The constancy of seawater composition and the ability to

determine chlorinity make measuring the composition of seawater much easier. If chlorinity is measured, the total salinity can be determined accurately. Not only that, but, if we know the chlorinity, we can find out how much sodium or magnesium or any other major component is present in any sample of seawater.

Today there is an even more accurate way to determine the salinity of seawater. If we know the temperature of the water, we can measure how well it conducts electricity, and from this we can determine the salinity since electrical conductivity in water depends on salinity and temperature. Not only is this method more accurate, but the instrument used to measure electrical conductivity can be lowered into the ocean to any depth. However, if we use the chlorinity method, a sample from a depth in the ocean must be brought up to the ship and analyzed. This is time-consuming and decreases the

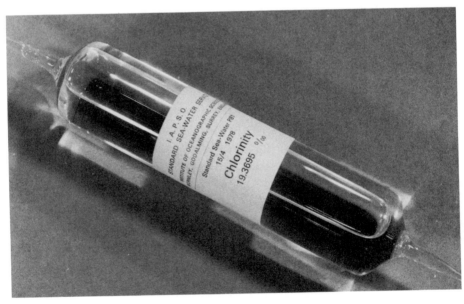

Standard seawater is sealed in glass ampules and sent to oceanographic laboratories around the world. The chlorinity of standard seawater has been accurately determined to four decimal places.

accuracy of the measurement. By measuring the electrical conductivity at some ocean depth, we can make a faster and more accurate analysis of salinity.

Why is it so important to accurately determine salinity? Actually it is extremely important. The reason is an accurate measurement of salinity will help us to accurately determine density. By determining density to three or four decimal places, we can understand the movements of water in the ocean.

Average seawater is only 2.5 percent more dense than fresh water. That is not much of a difference when you consider that ice is 9 percent less dense than fresh water. Differences in density within the ocean are even smaller than those between fresh water and seawater. Differences in density of 0.1 percent are common in the ocean. Such small differences are enough to cause the water in the ocean to move.

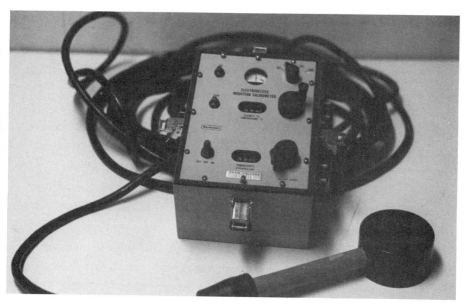

Modern determinations of the salinity of seawater are made by measuring its electrical conductivity with a salinometer. This salinometer is designed for use in water depths of less than 50 feet (15 meters).

If water of a slightly higher density is on top of water of a slightly lower density, there will be an unstable condition. The higher density water will sink to a position below water of lower density. The whole ocean is layered according to density.

Density, the mass of a volume of a given substance, is easy to determine in the laboratory but difficult to measure in the deep ocean.The density of seawater depends on its salinity and temperature. The higher the salinity and the lower the temperature, the greater the density.

The temperature is what makes it difficult to make accurate measurements in the deep ocean. If we bring a water sample from some depth up to the ship, its temperature will change and, therefore, so will its density. If, however, we can measure the temperature at depth in the ocean and also determine its salinity at that depth, we can calculate the density. Knudsen was the person who figured out how to do this.

If an area of the ocean becomes more dense, say because some of the water evaporates or cools, the water will begin to move in relation to the water around it. These density differences cause ocean currents.

Although the measurement of salinity is important in oceanography, it also gives rise to some interesting questions. If the major components are constant in their relationship to one another, what about the other elements dissolved in the water? How does an element, or, more accurately, the atom of an element, get into the ocean in the first place? Once an atom gets into the ocean, does it stay there forever, or does it stay only a short time? If we can understand why the salinity of the ocean behaves the way it does, can this tell us something about the way the rest of the ocean works? These are some of the interesting questions that must be answered. But first, there are even more basic questions: Why is there an ocean, and why is it salty?

# 3

# The Origin of Seawater

The earth is fascinating. Look at a globe and you quickly realize that most of the surface is covered by oceans. Nowhere else in our solar system is there a planet with so much water. The entire existence of all the complex life on earth depends upon water. Where did it come from? Why do we have oceans while the other planets do not?

You might think that all the water that exists on the earth came from the atmosphere in the form of rain. But if the whole atmosphere were to hold as much water as it possibly could and it all came falling to the earth as rain, it would raise the level of the ocean by only a few inches. Certainly the atmosphere could not be the source of the ocean. However, the fact that the atmosphere is made up of gases does provide us with a clue.

The gases in the atmosphere are there because the gravitational attraction of the planet keeps them there, just as it keeps our feet firmly planted on the solid earth. Astronauts can travel to the moon and spaceships to other planets only if they can get going fast enough to overcome this gravitational attraction and escape from the pull of the earth. We say that they must reach escape velocity. The same principle is true for gases.

If a gas molecule can get going fast enough, it can reach escape velocity and leave the earth. The way we can get a gas molecule to move faster is to heat it up. The higher the temperature, the faster the gas will move. The speed of the gas also depends on its mass. The heavier the gas, the higher the temperature must be before the gas will reach escape velocity. The lightest gases are hydrogen and helium. At the present average temperature of the earth, these gases are so light that they reach escape velocity and move out into space. We can not keep hydrogen and helium in the atmosphere. To keep these two gases, a planet must be either very cold or have a much stronger gravitational attraction than the earth.

We can get an idea about the history of the atmosphere by

M101, NGC 5457, a spiral galaxy 15 million light years away in the constellation Ursa Major. At the time of its formation, the ratio of elements in the earth should have been the same as those in the stars.

looking at the concentration of what are called the noble gases. These are helium, neon, argon, krypton, and xenon. These gases, unlike hydrogen or water vapor, for instance, do not combine chemically in nature. They always remain in gaseous form and do not become a part of the solid earth. Astronomers tell us that the earth accumulated from the material of stars. If this is so, the amount of noble gases present in the original earth was then probably in about the same ratio to other elements as they now are in the stars.

If we look at the present ratio of noble gases to other elements, we find that the earth has less of these gases than probably existed on the original earth. In fact, the earth has a greater shortage of the light noble gases than the heavy ones. The heavier the noble gas, the closer its concentration is to what we would expect. Since gases can escape from the earth by being heated, and the lighter gases would reach escape velocity sooner, this may suggest that the earth was much hotter in its early history than it is today.

If the earth lost much of its noble gases to outer space during its early history, it must also have lost those gases that make up most of the atmosphere. Water vapor, oxygen, and nitrogen are all so light that they could not have existed in the early atmosphere of the earth. If water vapor did not exist in the atmosphere, the oceans could not have existed either. Therefore, the atmosphere and the oceans must have accumulated on the earth after it cooled down enough to retain water vapor and the other gases.

If the atmosphere and the oceans were not originally present on the earth, they must have accumulated from gases released from the solid earth itself. Geologists have determined that the crust of the earth could not have held enough water to account for all that is in the oceans. Therefore, water and other gases must have come from the interior of the earth.

They would have to be the result of gas rising from the mantle, through the crust, and being released into the atmosphere. This is a process called outgassing of the mantle.

How does this gas get out of the interior of the earth? Actually, we see evidence of outgassing in many places. Volcanoes, hot springs, and geysers bring some water vapor and other gases from the interior to the surface. Measurements show that at the present rate of outgassing, enough gas would be brought to the surface over the billions of years of geologic time that have passed, to produce the oceans and the atmosphere.

STRUCTURE OF THE EARTH

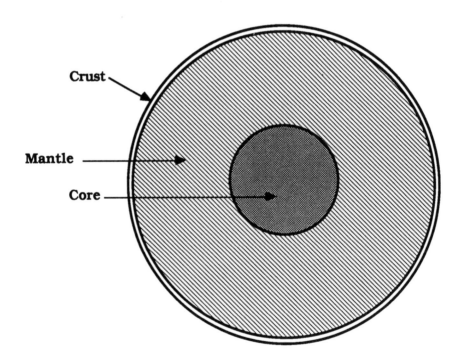

A section through the earth showing the structure of the interior. The band representing the crust is exaggerated for clarity.

But what about the other planets? Most scientists now believe that the other planets are no different from the earth. It is just that the rate of outgassing has been slower on the other earthlike planets, such as Venus, Mercury, and Mars. These planets are smaller, and possibly their interiors are cooler and have not outgassed as rapidly. Space probes have found volcanic features on these planets, indicating that the outgassing process has gone on.

If outgassing produced the oceans, where did the salts in the oceans come from? A logical idea would be that they are the products of the breakdown of rocks, which the rivers have

Steamboat Geyser in Norris Geyser Basin, Yellowstone National Park. The height of this eruption was estimated to be about 300 feet (91 meters) and may contribute to the outgassing process.

carried from the continents to the sea. If we compare the ratios of elements in average river water to those in average seawater, we find that they are not the same, but this does not mean that rivers do not carry elements to the sea. Many elements are used by organisms in the sea, thus changing the ratios of these elements to others. For example, many organisms build calcium carbonate shells. This removes some of the calcium from the ocean.

Rather than compare river water to seawater, a better way to approach this question might be to look at the composition of the continents, which have been weathered, and determine where the products of weathering might now be. We could set up an equilibrium balance, much like one used in science classes. The material weathered from the continents would be placed on the right-hand side of the balance, and the quantities of the same elements in the oceans would be put on the left-hand side of the balance. So, on our balance for the ocean, we would say that the amount of a particular element that has been weathered from the original continental rocks should balance with the amount of that element contained in the oceans, atmosphere, and sediments (those products of weathering that remain in solid form). If the two are not equal, then either our estimate of how much material has been weathered from the continents is too high or some of the elements in the ocean have come from some other source.

Any answers that might be obtained from this equilibrium would have to be approximate. We can determine the composition of the oceans, atmosphere, and sediments without too much trouble. The composition of the original continents, however, is a bit more difficult to determine. Samples from deep within the earth's crust must be brought up to the surface by drilling, and analyzed. Deep samples must be used so that we can be sure they have not been weathered.

Even with the problems in getting good estimates for the equilibrium balance, this system works quite well. There is a group of fairly abundant elements that is found in the samples from deep within the earth. When we find the differences in the ratios of elements in the samples and the present day surface, we get a figure telling us how much material has been weathered from the continents. Putting this amount of material on our balance opposite the quantities of these same elements now present in seawater, we find that they are equal. Therefore, this group of elements, which includes calcium, sodium, magnesium, and potassium, owes its origin to the weathering of the crust.

There is a second group of elements, which is not very abundant, that does not balance as well. However, this second group consists of elements that are so rare it is difficult to come up with an accurate estimate of how much there is in

A coring device being prepared for its descent to the seafloor. This device brings back samples that allow scientists to look at the history of the earth as it is recorded in the materials of the seafloor.

the ocean. This is probably the reason why they don't balance, so we really don't have to worry about them.

A third group of substances, however, does give us some concern, because it does not balance with the current quantities in seawater. This group consists of elements such as chlorine, bromine, sulfur, nitrogen, carbon, and the compound water itself. These substances are quite abundant so that errors in the estimates could not be the reason why they do not balance. Their abundance is far greater in the oceans than it should be if they came from weathering of the continents.

All the elements in this third group have one thing in common: they can exist as gases. Chlorine, bromine, and nitrogen can occur naturally as gases; sulfur, as sulfur dioxide gas or hydrogen sulfide gas; carbon, as carbon dioxide gas; and water, of course, as water vapor. Anything that can

Cable used in studies involving the ways sound waves react with the seafloor (seismic studies) aboard the *RV/Conrad* of Lamont-Doherty Geological Observatory. Seismic studies can determine what is below the surface of the seafloor and can look at the seafloor as it was in the past.

readily form a gas is called a volatile substance. Since these gases are more abundant in the ocean than they were on the continents' original surfaces, we call them the excess volatiles.

This inequality on our balance gives us a clue to the origin of seawater salinity. If we look at each of these gases we find that only 1 percent of the water vapor, carbon, and nitrogen that exists in the oceans could have come from the weathering of the continents. Only 2 percent of the chlorine and 20 percent of the sulfur could have come from weathering. There must be an explanation for such large differences between what could be produced from the continents and what actually exists.

The excess volatiles have the same origin as the oceans: outgassing of the mantle. The volcanoes, hot springs, and geysers, which carry water from the mantle to the surface, also carry the other excess volatiles. If we analyze the gases

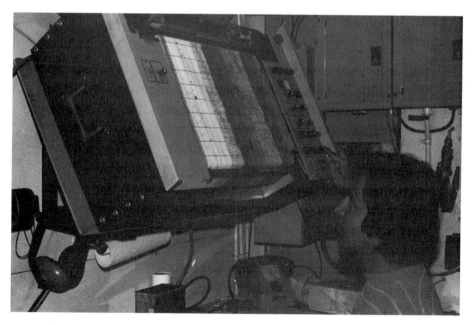

The electronic laboratory aboard the *R/V Conrad*. The results of seismic studies are recorded for future investigation.

coming out of the volcanoes, we find that the excess volatiles are there in high enough concentrations to explain why the equilibrium balance doesn't balance. Therefore, the interior of the earth is the source of some of the elements that exist in the ocean.

The origin of the salts dissolved in the ocean is not the same for every element composing those salts. The two most common elements dissolved in the sea, sodium and chlorine, come from different sources. Sodium comes from the weathering of continental rocks, whereas chlorine comes from the outgassing of the mantle.

The presence of the excess volatiles gives us the answer to another question. Was the sea always salty, or was the ocean originally fresh water that became salty over the length of time it has been in existence? Since both the water and the chlorine come from outgassing of the mantle, the first drop of water that condensed in the early atmosphere and fell to the earth probably contained chlorine and already had a large part of its salinity. When that first drop hit the solid earth it began the weathering process. This means that as water accumulated in the early ocean, it probably was almost as salty as it is now. Therefore, the salinity of seawater has probably not changed very much during all of geologic time.

Although the salinity of seawater has been fairly constant, this does not mean that there have been no changes in its composition over geologic time. Once life developed in the oceans, the minor elements in the ocean were affected. Photosynthesis releases free oxygen, which is dissolved in seawater. The cycle of life in the sea has a dramatic effect on several of the minor elements, which do not occur in constant ratio to the major elements. The variance in the concentration of these elements provides us with clues about some of the processes going on in the sea.

# 4

# Seawater:
# All Is Not Constant

An oceanographic research vessel cruises the Pacific Ocean, studying the composition of ocean water. Part of the process is like trying to find out what is at the bottom of a big glass of liquid without being able to look at it. The deep waters of the Pacific lie 5 miles (8 kilometers) below the rocking ship from which sampling bottles are painstakingly lowered. At the same time, water is collected from the surface to be compared to what is brought back from the bottom.

When these water samples are brought into the ship's laboratory and analyzed, the relative concentrations of the major components turn out to be constant, as expected. However, when the minor components are analyzed, relatively large differences are found between those in the surface water and those in the deep water. The concentration of phosphorus, nitrogen, and silicon is much greater in the deep water than in the surface water. In fact, these elements are almost absent at the surface. The concentration of carbon dioxide is also much greater in the deep water, but that of oxygen is much lower in the deep water than at the surface.

An electronic conductivity-temperature-depth (CTD) instrument being lowered overboard. The tubes at the top of the instrument are water sampling bottles that can be activated from the ship.

Just as it is fortunate for studies of the ocean that the major components of seawater occur in constant ratio, it is also fortunate that the minor components do not. The differences in concentration of the minor components provide oceanographers with a tool for understanding the biological, sedimentary, and mixing processes that occur within the ocean. In order to use this powerful tool, we must understand what controls the differences in concentration of the minor components.

Perhaps a look at the organisms living in seawater will give us a clue to the processes that take place there. What do these organisms eat? Animals eat plants and other animals. Plants, for the most part, gain their nourishment from substances dissolved in the water. Plants extract the elements carbon, phosphorus, and nitrogen from seawater to produce living tissue. Therefore, the growth of plants in the ocean is called primary productivity, since they need only the elements in seawater to produce tissue.

The plants must also use energy. This is provided by sunlight. Thus, plants can live only in surface waters of the ocean, where enough sunlight penetrates to allow tissue to be produced.

If we think about the vast expanse of the surface ocean and its tremendous depth, it is obvious that the plants responsible for primary productivity do not have roots penetrating the material of the bottom, as most land plants do. Oceanic plants are extremely small organisms that float in the water and go wherever the ocean currents take them. They extract carbon, phosphorus, and nitrogen from the water to make their tissue. As long as the water contains these elements, called nutrients, the plants will continue to flourish and multiply.

Much like the human beings who inhabit the land surface,

the plants of the ocean tend to be too productive for their own good. Although there is a fair amount of carbon in the water, in the form of carbon dioxide, there is not very much phosphorus and nitrogen. In their endless search for increased productivity, plants tend to remove all the available phosphorus and nitrogen. If this occurs, primary productivity ceases, and the only processes going on are the natural death of the plants and their consumption by animals.

As the plants die and primary productivity decreases, concentrations of phosphorus and nitrogen increase in the surface waters because new water replaces the old, nutrient-depleted water. This increase in nutrients spurs another round of primary productivity. The only thing that interrupts this continuous cycle is the cold weather of winter. Plants that live in warm water do not flourish in cold water.

The furious pace of productivity in the surface waters

Photograph taken from the *Alvin* of organisms on the East Pacific Rise referred to as "bean sprouts" (circular organisms). Perhaps our alteration of the ocean may affect this life on the midocean ridge.

leaves those waters with low concentrations of nutrients. But if the ocean waters circulate and new water replaces the old surface water, where does this new water get its nutrients? The answer is that the new water is actually water from the deep ocean, which rises to replace water at the surface.

Far below the ocean surface, the deep water moves slowly through the ocean basins and eventually reaches a point where it will rise to the sunlit upper layers of the sea. This is not its first trip to the ocean surface; it is estimated that the ocean waters complete a circulation every thousand years. Doesn't this present a problem? If the deep water was once at the surface, shouldn't all its nutrients be gone? How can it replace nutrients that have been used up by the plants? The key lies in what happens to the plants after they have been eaten or die.

Gravity works in the ocean as well as everywhere else. If a plant dies in surface waters, gravity causes it to sink. What if a plant doesn't die naturally but is eaten by an animal? Most of the animals that eat these plants are microscopic, just like the plants. But their digestive processes work much like ours. They extract what they can use and excrete the rest. Only about 10 percent of the tissue eaten by these microscopic animals is actually used in building their own tissue. Much of the rest is used for energy, and the unused elements are excreted. The material that is excreted is influenced by gravity just as are plants that die naturally.

The deep waters of the ocean are constantly bombarded by tiny particles of organic tissue falling through the water. It is much like a continuous rain of solid particles. These particles serve as food for another group of organisms, bacteria. The bacteria decompose the organic particles, releasing their elements back to the ocean water.

That is the answer! Nutrients that are locked in plant

tissue at the surface, are returned by bacteria to the water at depth. Therefore, deep ocean water contains the nutrients that have been removed from the surface. The rising of this water to the surface replaces the nutrients that were removed by plants hundreds of years before.

With this cycle occurring, it is not unreasonable to expect deep water to have higher concentrations of nutrients than surface water. The element silicon has a distribution similar to that of the nutrients, and its distribution can be explained by the same cycle. The most abundant plants in the ocean surface build tiny skeletons of silicon dioxide. As the plants extract nutrients from the water, they also extract silicon.

As long as the plants are alive, they protect their skeletons from dissolving. However, once a plant dies and the skeleton starts to sink, it starts to dissolve. The dissolving skeleton returns silicon to the deep water, just as the decom-

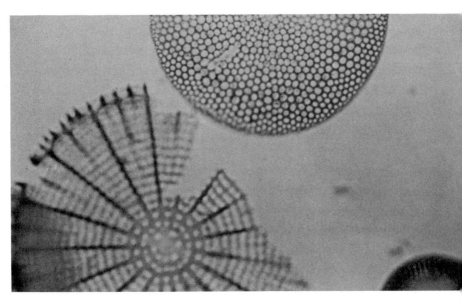

Diatoms, single-celled plants that construct a skeleton of silicon dioxide. They are responsible for most of the primary productivity in the ocean.

position of organic tissue returns nutrients. So the distribution of silicon in the ocean looks very much like the distribution of the nutrients. In fact, it may be that the element that actually limits the growth of plants in surface water is silicon. There is very little silicon in surface water, and it tends to be used up before the nutrients. Therefore, primary productivity may stop, temporarily, when the plants no longer have silicon with which to build their skeletons.

The distribution of carbon dioxide and oxygen is a little more complicated. Plants use carbon dioxide to make their tissue, producing oxygen. We would expect the concentration of carbon dioxide to be low in the surface water while the amount of oxygen would be high, which is what we find. However, more than that is happening.

Oxygen and carbon dioxide are gases. Their molecules can move back and forth between the ocean surface and the atmosphere. So there is a continuous exchange of these gases at the ocean surface. In addition, oxygen is used by animals in the deeper waters for respiration, and by bacteria for decomposition of organic tissue. So more than one process controls the distribution of oxygen. The concentration of oxygen at any place in the ocean must be a result of the balance among all the processes that use or produce it.

The controls of the distribution of carbon dioxide are even more complicated. It is used by plants but produced by animals during respiration. In addition, carbon dioxide is part of a chemical system that is used by animals. Many of the microscopic animals that feed on plants make their skeletons of calcium carbonate. A component of calcium carbonate ($CaCO_3$) is carbon dioxide ($CO_2$). Therefore, by making their shells, the animals cause changes in the carbon dioxide concentration in the water. The calcium carbonate shells dissolve in the deep ocean after the animals die, just as the plant skeletons do. The result is that carbon dioxide concentrations are low in the surface waters and higher in the deep ocean.

The ocean is like a huge kettle of chemicals that is being constantly stirred. Organisms prevent the elements they use from being mixed up completely by controlling their distribution. There are reasons why certain parts of the ocean have certain concentrations of some elements. But we know that elements are constantly being introduced from the land, coming to the ocean in rivers—continuously adding more chemicals to our kettle. If all that occurs is the stirring of the kettle, then the concentration of all elements in the ocean must continually increase, because we are adding and not taking away. We don't observe that happening. So what happens to the elements once they reach the ocean? That question may lead to an even better understanding of some of the processes occurring in the ocean.

Foraminifera, single-celled animals that construct a skeleton of calcium carbonate.

# 5

# The Ocean:
# A House Is Not a Home

One of the most popular places to visit in an amusement park is the fun house, a place of funny mirrors and scary rooms. You walk through it, laughing in one room and screaming in another. It is a great place to visit, but no one lives there. Perhaps that is why it is called a fun house and not a fun home. A home is a place where people live, not just a place to visit. It is a permanent residence; one doesn't just pass through laughing and screaming.

We can look at the ocean and ask whether it is a house or a home for chemical elements. Do these components enter the ocean and remain there forever? Do they enter the ocean and merely pass through on their way to someplace else? If they merely visit, how long do they stay? Does every element visit for the same length of time? These are questions we can answer for a great many of the elements that exist in the ocean, but not for all of them. There is still much for us to learn about the salts in the sea.

If we survey all the major rivers of the world, we find that each year, approximately $1.0 \times 10^{17}$ pounds ($4.6 \times 10^{16}$ kilograms) of water enters the ocean, which is more than the mass of all the living things on earth. The total mass of water in all the ocean of the world is $3.1 \times 10^{21}$ pounds ($1.4 \times 10^{21}$ kilograms). If we divide the total mass of the ocean by the mass of water entering it each year, we can determine how long it would take to fill up the ocean basins if they were empty when we started. The result of this division is that it would take slightly over 30,000 years to fill the ocean basins with the water that comes from these rivers. If the water in the ocean remained there permanently, the ocean would double its present size every 30,000 years. Since the ocean has been there for hundreds of millions of years, it is clear that the water must not remain permanently in the ocean, but merely passes through. As the total mass of water in the ocean probably has not changed much over millions of years, we say that the ocean is in a steady state. That means the water must leave the ocean at the same rate that it enters; otherwise the mass of the ocean would change.

The primary way by which water leaves the ocean is through evaporation from the sea surface to the atmosphere. Since the water does not remain permanently in the ocean, there is a definite period of time during which it will stay there, called its residence time. The residence time of anything is the time it takes to replace all that is present with new material. The division of the mass of the ocean by the mass of water entering the ocean each year is a calculation of residence time. Therefore, the residence time of water in the world ocean is somewhat over 30,000 years.

Although water can be removed from the ocean by evaporation, the salts dissolved in the water remain behind. If there were not some mechanism to remove these salts, the ocean would continually get saltier. As with the water, the salts in the ocean are also in a steady state. The average salinity of the ocean does not change over long periods of time. Therefore, there must be a mechanism for the removal of salts from the ocean.

There are several ways in which salts can be removed. The simplest occurs as waves break, either at the shoreline or as whitecaps in the open ocean. The spray is thrown into the air, where it evaporates, leaving tiny particles of airborne salts. Anyone who has been at the beach on a windy day has acquired a layer of tiny salt particles from the air on his skin. Such particles remain in the atmosphere for days or weeks before raindrops dissolve them and bring them to the ground. They can fall to the ground far from the ocean and be removed from their house.

Even as organisms control the distribution of some elements in the ocean, they also cause the removal of these substances. Not all the organic material that falls from the surface

A walk along the ocean's edge permits people to carry away with them some of the salt of the sea. Evaporation of sea spray leaves a coating of salt on one's skin.

of the ocean is decomposed, with its elements being returned to the water. A small fraction of the material settling through the water actually reaches the bottom and is buried in sediment. The amount that becomes buried depends on how the element is used by organisms. In the case of phosphorus, which is in the soft tissue of organisms, approximately 1 percent of that which falls through the water is buried in sediment. If the only way that phosphorus is removed from the ocean were by burying organic tissue, then its residence time would be 105,000 years. However, if we divide the total mass of phosphorus in the ocean by the amount entering it each year, the residence time is 69,000 years.

The difference between the two figures must mean that there is another way for phosphorus to leave the ocean, since a shorter residence time would mean a faster removal. Oceanographic research ships dredging the sea floor have found sev-

Manganese nodules on the seafloor. The formation of these nodules composed of iron, manganese, and many other substances removes elements from seawater.

eral areas where the bottom is littered with potato-sized masses, called phosphate nodules, that have precipitated from seawater and contain a high concentration of phosphorus. It seems that the calculations were correct. There is more than one way for phosphorus to be removed from the ocean.

As we have seen, elements like silicon and calcium go into the shells of the organisms. Shells that fall through the water should be destroyed at a slower rate than soft tissue. The amount of the element that is buried in the sediment depends on how fast the shell will dissolve in seawater.

Shells composed of silicon dioxide dissolve rapidly. Still, approximately 5 percent are buried in the sediment, and their silicon is thus removed from the ocean. If that is the only way that silicon is removed from the ocean, its residence time is 20,000 years. Calculation of silicon residence time, using the mass in the ocean and the river input, also yields a residence time of 20,000 years. Comparing the two methods provides us with a way of checking to make sure we have the correct removal mechanism. It is evident that the most important mechanism for removing silicon from the ocean is burial of shells in the sediment.

Shells made of calcium carbonate do not dissolve as readily as those of silicon dioxide. Fourteen percent of calcium carbonate shells become buried in the sediment of the seafloor. If this were the only removal mechanism, the residence time of calcium would be 800,000 years. But if we calculate residence time by the other method, we come up with a value of 1,100,000 years. Hold everything! This time, calculation by the second method produced a larger number. If a smaller number meant that phosphorus was being removed by a mechanism other than organisms, perhaps a larger number means that input from streams is not the only source of calcium. What is that supply? Where could the calcium be coming from?

*February, 1979. The Pacific Ocean south of Baja California.* The deep-diving submarine *Alvin* with its three-man crew is lowered from the deck of the mother ship and begins a two-hour descent to the East Pacific Rise, 1½ miles (2.5 kilometers) below. The *Alvin's* running lights are turned on to provide some illumination in the total darkness of the seafloor, and a routine survey begins.

Suddenly an unexpected sight becomes apparent at the edge of the lighted area: a structure that resembles a chimney. Spewing from the chimney is what looks like black smoke. In reality it is hot water mixed with metal sulfides that give it a black color. After several attempts, the water temperature is measured. It is 350°C (660°F). From the water's composition, it is evident that the smoky plume is seawater that has circulated through the hot rocks of the midocean ridge system before reappearing on the seafloor. The chimneys spewing the

The deep-diving submersible *Alvin* being prepared to take its crew to the depths of the ocean.

blackened water have been termed black smokers. Since that initial dive of the *Alvin*, black smokers have been found in several portions of the midocean ridge system.

Analyses of the waters coming from the black smokers show that they are different from the bottom waters of the ocean. As the ocean waters circulate through the rocks of the midocean ridge, chemical reactions occur between the water and the rocks. One result is that calcium is extracted from the rocks and added to the water. The source of the missing calcium has been found.

Just as the source of some of the calcium in the ocean was a mystery before the black smokers were discovered, there was also a question about magnesium and sulfur. The question was not where do they come from, but rather where do they go? There were problems in attempting to explain how these elements were removed from the ocean. Black smokers

An active black smoker photographed on the East Pacific Rise in 1979. Note the chimneylike feature from which the black water is coming.

now provide the answer. As the water circulates through the midocean ridge system, magnesium and sulfur are removed, and the water coming from the black smokers contains a smaller amount of those elements than normal seawater.

The trip through the fun house that we call the ocean is not the same for all elements. There are many entrances to the house and many exits. Any one element can get in and leave only through a limited number of ways. Traffic is flowing in a very complex manner that controls the total amount of an element in the ocean.

If the rate at which an element enters or leaves the ocean changes, then the total amount of that element in the ocean will vary. Eventually, a steady state will be established and the amount entering in a given time will equal the amount leaving in that time.

That brings us back to the original question. Mankind is continuously dumping things into the ocean, and, for that matter, into all the waters of the earth. We are adding gases to the atmosphere. We are using huge quantities of water. Can mankind affect the composition of the oceans enough to cause a measurable change?

# 6

# Prospects for the Future

Did you ever think carefully about your home? You would like it to be warm in the winter and cool in the summer. That requires energy. Energy to heat your home and energy to run a fan or air conditioner to cool it. If you are hungry, you may go to the refrigerator to look for something to eat. The refrigerator requires energy to operate. Whatever you choose to eat has to be raised or caught, which requires energy. If what you select must be cooked, you turn on the stove. That requires energy. Perhaps there is a car parked in front of your home. It requires energy to operate. The clothes you wear require the use of energy to make them. Almost anything you can think of requires energy to produce. The world has evolved into a society that depends upon energy production.

Although it is difficult to get an accurate estimate, there may be about 7 billion people in the world by the year 2000, all of them requiring the use of energy, even if it is produced by burning wood to keep warm and cook food. Energy production releases gases and particles into the atmosphere. In the United States alone the gases and particles released into the atmosphere may be about 1 billion tons each year, or about 3.5 tons for every person in the country. If an average person weighs about 150 pounds, then 47 times his own weight is released into the atmosphere each year.

One of the consequences of energy production is the release of carbon dioxide into the atmosphere. As we have seen, this gas can move between the atmosphere and the ocean. As its concentration in the atmosphere increases, about half of the added carbon dioxide will move into the oceans. If the total mass of carbon dioxide increases because more is coming into the ocean, then it must begin to leave the ocean at a faster rate than at present. It cannot be used by plants during photosynthesis, because as we have seen, the limiting element for plant growth in the ocean is silicon. As long as the silicon supply remains constant, the microscopic plants in the surface waters will not absorb the added carbon dioxide.

The most likely mechanism for carbon dioxide removal appears to be the dissolving of calcium carbonate from the sediments of the seafloor. That will convert the carbon dioxide to the bicarbonate ion and release calcium to the oceans. This, of course, will throw the calcium steady state out of whack, and there will have to be chemical adjustments to return to it.

It is evident that the large-scale experiment that mankind is conducting will have a great effect on the composition of the ocean. We cannot affect only one element in the ocean. Everything depends on everything else. If we change the carbon dioxide content, the calcium will be affected. If we change the calcium content, magnesium and sulfur will be affected. And on, and on, and on.

Suppose we were to stop the experiment today. As of this moment, mankind would no longer add carbon dioxide to the atmosphere. Would that halt any changes that might occur? It has been estimated that it would take the surface waters of the ocean one year to achieve a steady state situation with the atmosphere. That is a fairly fast response. However, the extra carbon dioxide must leave the system by reacting with calcium carbonate on the seafloor. Therefore, it must work its way to the deep sea.

The time required for the waters of the deep ocean to return to a steady state situation with the atmosphere is on the order of hundreds of years. Once that occurs, the calcium carbonate in the sediment must begin to dissolve and reach a

The "Pompeii worm" forms a tube from minerals in the hot water from the vents on the East Pacific Rise and cements itself near the chimneys that spew solutions hotter than 350°C (660°F). Mankind's activities may affect the fate of these organisms.

steady state situation with the deep ocean waters. That would occur on a time scale of thousands of years.

The problem has still not been resolved. There would then exist a calcium problem. The excess calcium from the dissolving of calcium carbonate would have to be taken care of. To do that would take several tens of thousands of years. It becomes clear that the effect of mankind's energy production on the oceans of the world will be great and will last for tens of thousands of years, even if we stop all release of carbon dioxide into the atmosphere. We are truly in the midst of a great experiment, one we cannot stop even if we want to.

Mount St. Helens as viewed from the northwest on July 22, 1980, about one minute after the beginning of the second eruptive phase. Highest point of the eruption cloud is about 9,900 feet (3,000 meters) above the highest visible point on the rim of the volcano. Volcanic activity is a part of the outgassing process.

Just as the world depends upon energy production, it also depends upon the chemical industry. Not only do the clothes we wear require energy to manufacture, they are also probably made of chemicals that do not occur naturally. Most of the items we use in our everyday life are made from chemicals created in a laboratory. In the manufacture of these chemicals there are waste products. When the items we use wear out, we dispose of them. Much of the waste winds up in the ocean.

The release of carbon dioxide is producing a permanent change in the ocean, and so too is the disposal of waste. Once

An oceanographic research vessel, the *R/V Conrad*, at sea. All studies of the deep ocean must be done from such vessels.

these substances are placed in the ocean, there is no way to retrieve them. The great experiments that are going on are permanent. We *cannot* return the ocean to its original condition.

Since we have already altered the ocean, every effort must now be made to insure that we minimize any further harm to the ocean system. To do this, society must adopt an entirely different way of thinking. It is not realistic to give up our technology and return to a primitive way of life, but we can work to make our technology more efficient. If our use of energy were more efficient, this would at least decrease the

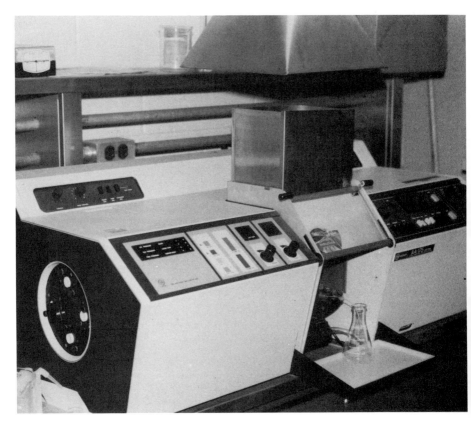

An atomic absorption spectrometer. Instruments such as this can detect small, yet toxic, quantities of pollutants in seawater.

amount of energy that would be needed. That may mean that many more energy-generating plants would not have to be constructed in the future. Alternate ways of producing energy must be devised, ways that do not release carbon dioxide and other gases into the atmosphere.

Just as we must change our thinking with regard to energy production and use, we must also change our thinking about the disposal of wastes. The earth can no longer tolerate mankind's heaping up garbage in landfills or dumping in the ocean. If wastes are to be produced, and there is no way to avoid producing them, they must be eliminated at the source of production. In most cases what are now discarded waste materials could be collected at the places where they are produced and recycled into productive materials. Such processes would, of course, often mean that goods would no longer be made as economically, but we may have to pay more for goods in order to survive.

Many organizations are actively pursuing alternatives to our present techniques of energy use and production, and waste disposal. Unfortunately, they are not all proceeding in a realistic manner. Alternatives to our present technologies must be based upon scientific facts, an understanding of the entire system, and the realities of human nature. There are no simple solutions. We must build a community of scientists in all disciplines and areas of employment who realize that no matter what mankind does, it will probably affect the ability of our planet to support life.

Seawater is salty. The substances that make up the salinity of the ocean are part of an interrelated system. The distribution of elements in the sea evolved to a steady-state relationship millions of years before mankind developed. Today we, in our short existence on earth, have reached the point where we can affect the oceans. We can cause, and have caused, disease and death by abusing the ocean. We are in the process of altering the total composition of the oceans for thousands of years to come. We must proceed with caution. Although we have some idea of how the ocean works chemically, we do not understand everything. We cannot predict what will happen when we unknowingly perform our great experiments. The ocean is vast, but its chemical system is in delicate balance, and we must be careful about upsetting that balance.

# Glossary

*Alvin*—The name of the deep-diving submersible vessel that took scientists down to observe the black smokers and fantastic organisms at the center of the midocean ridge.

**atom**—The smallest particle of an element that can combine with similar particles of other elements to produce compounds.

**bacteria**—Single-celled organisms that lack chlorophyll and an evident nucleus. Most bacteria are capable of decomposing organic matter; some cause disease.

**black smoker**—When the superheated water at the center of a midocean ridge rises to the surface, it contains many dissolved metals. Though exiting the seafloor clear, the water quickly cools and the dissolved metals form compounds, resulting in black, smokelike water.

**chlorinity**—The chloride content of seawater expressed in grams per kilogram (g/kg) or parts per thousand.

**compound**—A substance containing two or more elements combined in fixed proportions.

**continent**—About one third of the earth's surface that rises above the deep ocean floor to be exposed above sea level.

**continental shelf**—A gently sloping surface extending from the low-water line to a marked increase in slope around the margin of a continent.

**crust**—The thin cover of the earth above the mantle. In oceanic regions, the crust has an average thickness of about 6 miles (10 kilometers), and in continental areas it has an average thickness of 21 miles (35 kilometers).

**decomposition**—The breakdown of nonliving organic material primarily by bacteria that extract some of the products of decomposition for their own needs and make available the compounds needed for plant production.

**density**—The ratio of mass to volume for any material. Usually expressed in grams per cubic centimeter (g/cc). For ocean water with a salinity of 35 parts per thousand at 0°C (32°F), the density is 1.028 g/cc.

**diatoms**—A single-celled plant with two overlapping silicon dioxide shells.

**dredge**—A basketlike apparatus that is dragged along the ocean bottom in order to retrieve biological or geological specimens.

**East Pacific Rise**—The midocean ridge that runs north-south along the eastern side of the Pacific Ocean. It is the predominant location upon which the hot springs and black smokers have been discovered to date.

**element**—One of the approximately 106 chemically different fundamental constituents of matter, each of which is composed entirely of the same kind of atoms.

**escape velocity**—The velocity that must be reached in order to escape the gravitational attraction of the earth.

**evaporation**—The physical process of converting a liquid to a gas.

**excess volatiles**—Volatile compounds found in the oceans, sediments, and atmosphere in quantities greater than the chemical weathering of crystalline rock could produce. They are considered to be produced by outgassing of the earth's interior.

**foraminifera**—Single-celled floating and bottom-dwelling animals that possess protective coverings, usually composed of calcium carbonate.

**geyser**—A hot spring that periodically produces steam, which is then blown out at the surface of the earth.

**gravity**—A force with which two bodies pull on each other and which is in proportion to their masses.

**manganese nodules**—Rounded objects containing compounds of iron, manganese, copper, and nickel, found scattered over the ocean floor. They are formed around sharp or rough objects such as rock particles and sharks' teeth.

**mantle**—The zone between the core and crust of the earth. The mantle extends downward to a depth of about 1,740 miles (2,900 kilometers).

**mass**—The actual quantity of matter in any body.

**midocean ridge**—A long mountain range that extends through all the major oceans, rising from 0.6 to 1.8 miles (1 to 3 kilometers) above the deep ocean basins. Averaging 900 miles (1,500 kilometers) in width, it is a source of new oceanic crustal material.

**molecule**—The smallest particle of a compound or element that occurs as a combination of two atoms and, in a free state, retains the characteristics of the substance.

**noble gases**—Any of the six elements that have no tendency to react with any of the other elements in nature: helium, neon, argon, krypton, xenon, and radon. They are all gases under usual conditions.

**nutrients**—Any number of organic or inorganic compounds used by plants in primary production. Nitrogen and phosphorus compounds are important examples.

**outgassing**—The loss of gas from within a planet to space.

**phosphate nodule**—A black, gray, or brown rounded mass ranging in diameter from a fraction of an inch to more than 12 inches (30 centimeters), consisting of a nucleus of some naturally occurring material that is more or less enveloped in calcium phosphate.

**photosynthesis**—The process by which plants produce carbohydrate from carbon dioxide and water in the presence of chlorophyll, using light energy and releasing oxygen.

**primary productivity**—The amount of organic matter organisms make from inorganic substances within a given volume of water or habitat in a unit of time.

**radiolaria**—Single-celled floating and bottom-dwelling animals that possess protective coverings, usually composed of silicon dioxide.

**residence time**—The average length of time a particle of any substance spends in the ocean. It is calculated by dividing the total amount of the substance in the ocean by the rate at which it enters the ocean or the rate at which it leaves.

**respiration**—The process by which organisms use organic materials (food) as a source of energy. As the energy is released, oxygen is used, and carbon dioxide and water are produced.

**salinity**—A measure of the quantity of dissolved solids in ocean water. It is expressed in parts per thousand, which is the number of grams in 1,000 grams of seawater.

**salinometer**—A device that measures electrical conductivity and relates that to the salinity of ocean water to a precision of 0.003 parts per thousand.

**shoreline**—The line marking the intersection of the water surface with the shore. It moves up and down as the tide rises and falls.

**standard seawater**—Seawater that has been analyzed for chloride ion content to the nearest ten thousandth of a part per thousand. It is then sealed in ampules and labeled to be sent to laboratories throughout the world. All salinity analyses are made in reference to this standard so that they can be compared with one another all over the world.

**steady state**—A condition where a situation does not change and appears to have no beginning or end. In the case of the oceans, dissolved materials are constantly being added and removed, yet their total concentration does not change.

**volatile**—Evaporating rapidly, easily becoming a gas. Also, as a noun it means any volatile substance.

**volcano**—A vent in the surface of the earth through which magma and associated gases and ash erupt.

**volume**—The amount of space that an object or substance occupies.

**wave**—A disturbance that moves over or through a medium with a speed determined by the properties of the medium.

**weathering**—The destructive processes by which solid materials, on exposure to atmospheric agents at or near the earth's surface, are changed in color, texture, composition, firmness, or form, with little or no transport of the loosened or altered material.

# Further Reading

Bernard, Harold W. *Greenhouse Effect*. New York: Harper & Row, Publishers, Inc., 1981.

Blair, Carvel. *Exploring the Sea: Oceanography Today*. New York: Random House, Inc., 1986.

Carson, Rachel. *The Sea Around Us*. New York: Oxford University Press, Inc., 1951.

Cousteau, Jacques Yves. *Challenges of the Sea*. Mountain View, Calif.: World Publications, 1974.

Earle, Sylvia A. and Al Giddings. *Exploring the Deep Frontier: The Adventure of Man in the Sea*. Washington, D.C.: National Geographic Society, 1980.

Fine, John C. *Oceans in Peril*. New York: Macmillan Publishing Company, Inc., 1987.

Goldin, Augusta. *Oceans of Energy: Reservoir of Power for the Future*. New York: Harcourt Brace Jovanovich, Inc., 1982.

Hargreaves, Pat, editor. *Seas and Oceans*. Morristown, N.J.: Silver Burdett Company, 1981.

Lambert, David, and Anita McConnell. *Seas and Oceans*. New York: Facts on File, Inc., 1985.

Limburg, Peter. *Oceanographic Institutions: Science Studies the Sea.* New York: Elsevier/Nelson Books, 1979.

Parks, James. *A Day at the Bottom of the Sea.* New York: Charles Scribner's Sons, 1977.

Polking, Kirk. *Oceans of the World: Our Essential Resource.* New York: Philomel/Putnam Publishing Group, Inc., 1983.

Poynter, Margaret, and Donald Collins. *Under the High Seas: New Frontiers in Oceanography.* New York: Macmillan Publishing Company, Inc., 1983.

Vogt, Gregory. *Mars and the Inner Planets.* New York: Franklin Watts, Inc., 1982.

Whipple, A. B. C. *Restless Oceans.* Morristown, N.J.: Silver Burdett Company, 1983.

Wright, Thomas. *The Undersea World.* Morristown, N.J.: Silver, Burdett Company, 1982.

# Index

63